S0-DVG-008

FIREFIGHTERS

Written By:
Herbert I. Kavet

Illustrated By:
Martin Riskin

Manufactured in the United States of America

30 29 28 27 26 25 24 23 22 21 20 19 18 17 16 15 14 13 12 11 10 9 8 7 6 5 4 3 2 1

Ivory Tower Publishing Co., Inc.
125 Walnut Street
P.O. Box 9132
Watertown, MA 02272-9132
Telephone #: (617) 923-1111 Fax #: (617) 923-8839

F.B.I. WARNING

THE MATERIAL IN THIS BOOK IS PROTECTED BY COPYRIGHT LAWS OF THE UNITED STATES OF AMERICA AND OTHER COUNTRIES WHERE THEY DON'T EVEN SPEAK ENGLISH.

1. Don't forget to pay for this book. Even if the store is burning, you're not allowed to just stick it under your coat.
2. If the store is not burning, you can't just read all the cartoons and then put it back. Three or four pages are all that's allowed under Federal Law.
3. If you receive this book as a gift, you have to write an effusive thank you note.
4. You're not allowed to tell any of these jokes' punch lines to your buddies. They have to buy the book, too.
5. If there are any errors or omissions, complain to your Chief, not the FBI.

Introduction

Most often, it doesn't matter much what I write in INTRODUCTIONS since no one ever reads this stuff. Most people just want to get on to the cartoons. Given the choice of reading introductions or looking at cartoons, the cartoons win every time. But with firefighters, you can't be too careful. Sometimes firefighters have lots of time to sit around and read, and sometimes they get pretty desperate for something to look at, especially in the toilet. Lots of my books end up as reading material in toilets. So, if that's the case, I'd like to put something here, and what better place to thank Chief Murphy and all the guys at the Wayland Fire Department for their help, and I'm sure they've forgotten all about that little incident with the fireworks and the grass fire last July 4th. Also, many thanks to Leo Stapleton, whose books helped educate me about a firefighter's life as well as providing many choice four letter words to use in downtown Boston.

The Firefighter Gene

There is a certain mutant gene that resides in the brains of firefighters that must make them choose this dangerous line of work. Why else would grown men and women go through the training, spend the long boring hours waiting and then risk their lives to save other people's property. Ah, yes, there is an occasional beautiful maiden to rescue and that's pretty exciting, but those rescues are few and far between. It's obvious this firefighter gene exists, just look at the number of firefighters whose fathers were firefighters. Many think this gene is passed along by playing with unwashed firetrucks as a kid.

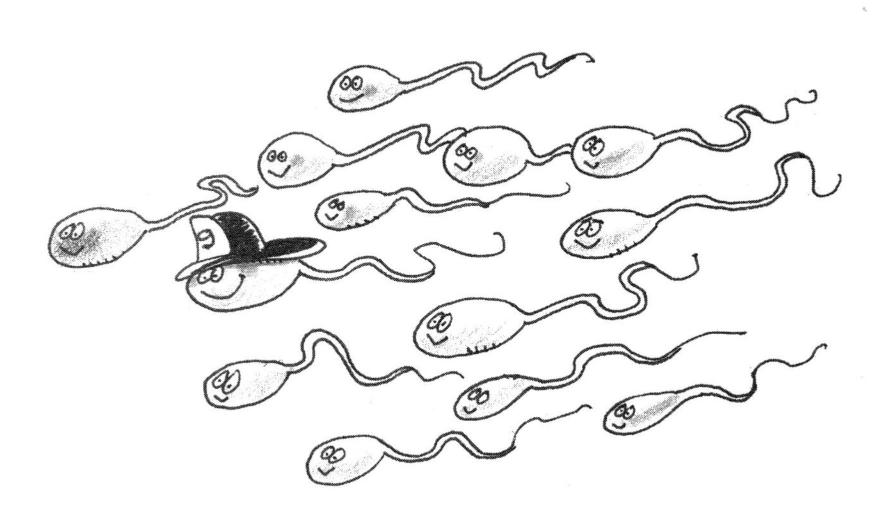

What's Going On Here

Fires are chaotic, tumultuous scenes, especially when you just arrive. What to do first? Hook up at a hydrant, force an entry, do some venting, initiate a search for victims, administer first aid, save some valuables, rescue a pet or perhaps tend to the buxom young woman who just escaped in only her underwear and is running around the fire engine trying to keep from getting chilled? Ha Ha. Trick question. The first officer to come on the scene is going to take care of her, you have to go put out the damn fire and now you'll have to wait for another officer to arrive to organize things.

Why Firefighters Are Irresistible To Women

Firefighters are brave and daring, generally handsome and fit, and lead exciting heroic lives. What more could any woman hope for? Women firefighters tend to frighten off all but the most secure men, and most end up marrying male firefighters. Recently, some fire departments have taken to designing a special pocket in their uniforms to enable firefighters to carry a small wooden club with which to fight off aggressive women. It's a fact that the only men more irresistible than firefighters are humor book writers.

Why Firefighters Are Irresistible

Firefighters

1. Are hung like a hose.
2. Aren't afraid of the heat.
3. Do it on their knees.
4. Always come in emergencies.

Cartoon Idea by Collie Grey

Female Firefighters

Notice the illustration on this page, of a vigilant and attractive female firefighter. Much of the copy and cartoons in this book are male oriented and, so that I don't get nasty letters from all those brave and skilled lady firefighters out there (like we do when we accidentally ship sex books to Mississippi), I wanted a page devoted fully to them. These strong and skilled women share the dangers with the guys, and don't squawk about sharing bathrooms or bunk rooms or breast feeding in a corner of the fire house.

Firefighters And Junk Mail

I'm sure you firefighters have suspected this all along. Firefighters get more junk mail than any other profession. Entire forests in Oregon are cut down each week just to print the catalogs and offers that firefighters receive. Badges and T-shirts and videos and pins and sweatshirts and model engines and coin banks, not to mention how to earn big money cleaning chimneys or stuffing envelopes or raising squirrels in your spare time. Every year thousands of firefighters are injured by the avalanche caused by opening their mail boxes too quickly when returning from vacation.

Electricity

Electricity starts lots of fires but no one really knows what electricity is. Once at a fire academy lecture, the teacher noticed one rookie dozing in the back. He startled him awake by calling on the slacker and asking, "Mr. Murphy, what is electricity?" Murphy hemmed and hawed and finally stuttered, "Well, Sir, I knew but I forgot." The teacher addressed the class solemnly, "Gentlemen, you are witnessing one of the great tragedies of science. Mr. Murphy here was the first person in the world to know what electricity is . . . and he forgot."

"HEY FELLAS, JAKE FOUND THE LINE!"

Cartoon Idea by Collie Grey

Electricity

While we don't really know what electricity is, we do know it's made up of amps and volts and stuff like that, and it travels around coils and conductors, passing through terminals and fuses in the form of single phase or three phase current, whatever that is. No one really knows what relays or solenoids do either. Mostly electricity just sits there and waits to get you when you are least expecting it. When working with potential electrical hazards, it's always safest to let someone else touch the stuff first.

"ATTA BOY, CHIEF! HOLD IT RIGHT THERE TILL WE GET PAST YA!"

Cartoon Idea by Collie Grey

Pagers

Lots of firefighters carry pagers or some kind of radio. These electronic conveniences used to impress people and were great fun to carry, but nowadays with every teenager and drug dealer beeping away, people tend to find them an annoyance rather than a status symbol. Pagers do, however, have a profound effect on a firefighter's sex life.

1. Pagers are responsible for the majority of divorces among firefighters.
2. Pagers are a cause of frigidity in 62% of firefighters' wives.
3. Pagers are an excuse for premature ejaculation in 47% of all firefighters (male).

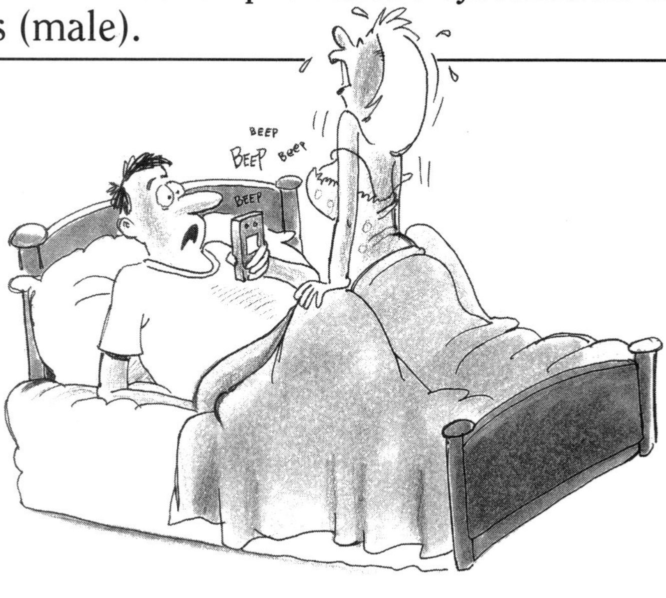

"HURRY UP, HON... I GOTTA GO!"

Saturday Morning At The Firehouse

This is the day that firefighters get to rescue all their handymen neighbors. They call them handymen because they lose their hands so often, mostly fingers actually, and get horrible lacerations over the rest of their bodies. The EMT's are busy all morning getting the paper pushers, who love to play carpenter and lumberjack, off the roofs and out from under cut trees.

"HOW LONG WILL IT BE BEFORE HE CAN FINISH THE GUTTERS?"

Firefighters At Parties

Firefighters tend to stick together at parties. They love to retell their favorite fire stories and love the camaraderie that comes to men and women who do a dangerous job together. People don't bug firefighters for advice like they do doctors and lawyers at parties. Even an unsolicited, "Say if you'd put a screen in front of that fireplace, your curtains wouldn't catch fire so often," draws a look like you left your pants open.

"I LIKE FRANK... HE ALWAYS KNOWS WHEN HE'S HAD ENOUGH TO DRINK."

Firefighters At Parties

A firefighter can always escape a dull party his wife has dragged him to by telling a horror fire story or two. It's a great way to end a party at your own home that's lingered too long, especially if you can get the guests outside before they throw up or faint. Firefighters just love to gross people out.

"WELL THAT CERTAINLY SOUNDS LIKE A HORRIBLE BURN."

Handling The Elderly

In every town, three or four elderly people will account for 80% of your emergency calls. These people are often lonely and scared, and will call for help when they fall or lock themselves out or smell smoke from a neighbor's grill. In my town, a woman called for help saying she'd fallen and couldn't get up. When the rescue squad arrived and started to enter the locked house by forcing a window, she jumped up, yelled at them for threatening to break her window and then asked them to help put away her groceries.

You Just Know It's Going To Be A Bad Day When...

1. You can't find the fire.
2. The hydrants are frozen.
3. All the women have been rescued by the time you arrive.
4. You arrive at the fire without your pants.
5. A state administrator arrives and announces he's there to help.

Cartoon Idea By Collie Grey

Animal Rescues

Giving mouth to mouth resuscitation to a Doberman, pulling a duck from a chimney or chasing a cat around a tree is not why most people become firefighters. But sometimes you feel you should help out, and more often, there is no one else to do the job. Even though it isn't your job, everyone is sort of staring at you, and since you haven't rescued a human being in so long, you go ahead and mostly are rewarded with a lot of scratches.

Cartoon Idea by Collie Grey

"LIKE YOU SAID CHIEF. WE TOLD HER TO JUST PUT OUT A BOWL OF MILK AND HER CAT WOULD COME DOWN WHEN IT WAS HUNGRY."

Cartoon Idea By Collie Grey

Gorilla In Tree Rescue

The fire station received a call from a frantic Mr. Fudsmith that he had a gorilla up a tree in his backyard. The dispatcher assured the man that their emergency gorilla rescue unit was on the way. Firefighter Tracey arrived in no time, but unfortunately his partner was on vacation. He quickly unloaded his gorilla kit, consisting of a ladder, club, rope, net, a fierce pit bulldog and shotgun. "Mr. Fudsmith," he said, "you can relax now. I've done this many times before and we'll have that gorilla out of your tree in no time. But my partner is away so you'll have to help. Now let me explain very carefully what's going to happen. First, I climb the ladder, sneak behind the gorilla and whack him with the club. The gorilla will lose his balance and fall out of the tree. This specially trained dog will rush up and bite his testicles. This causes the gorilla to faint, giving me a chance to drop the net on him, tie him up and take him to the zoo!" Fudsmith nodded that he understood. "But what is the shotgun for?" he asked. "Oh, yeah," said Tracey, "I almost forgot. You hold the shotgun. It's never happened before and it's very unlikely, but in case I slip and fall out of the tree—shoot that dog!"

Embarrassing Rescues

These are the ones you don't tell your wife about. Every firefighter has his favorite, often involving a woman stuck in a shower, toilet or tub, and don't worry guys, I'm not printing your secret story here. While you can't talk too much about these rescues at home, at least they enliven the conversation at the firehouse for years.

"OK, OK... STAND BACK, YOU'VE ALL SEEN A 400 LB. WOMAN STUCK IN A BATHTUB BEFORE."

Coffee

Know what fuels firefighters? Coffee—strong and hot and in large quantities. At most big fires, it's about as important as water. In winter, it's probably more important. Without lots of coffee, a department's manpower would have to be increased by about a third.

"HEY CHIEF! WE RAN OUT OF COFFEE... YA WANT A SODA POP?"

Cartoon Idea by Collie Grey

Coffee

Real firefighters don't drink cups of coffee, they drink mugs of coffee. The tougher the firefighter, the bigger their mug, and the really alert ones have mugs that hold 7 to 10 cups worth. These firefighters may be the most awake, but they still miss lots of fires 'cause they spend so much time going to the toilet.

Staying In Shape

It's important for firefighters to stay fit and strong. After all, this is a profession where wrestling a hose, climbing stairs with 60 lbs. of equipment and dashing up and down ladders rescuing beautiful women (and maybe men) is an everyday part of the job. There are two schools of thought in fitness. Those of the "No Pain, No Gain" school of thought, lift heavy weights, swim, row, ride prodigious distances, generally smell up health clubs and have hard, little asses. The "Can't Be Bad If It Feels Good" people hang out in the hot tubs, saunas and bars. They smell like roses, are generally cuddly and tend to have soft, droopy tushes that are comfy when watching TV.

Staying In Shape

Losing weight through exercise is easy and rewarding, provided you like to run, say, from Chicago to Cleveland each week. You need to burn 3,500 calories to lose one pound so if you're interested in dropping 10 or 15 pounds, you'll have to ride your bike across China, or swim to Haiti or something like that.

Actually, you <u>can</u> lose weight by exercising if you do a little each and every day, but the only people with the discipline to do this are already skinny.

Bud was sure he'd meet lots of women with his new running program.

Fear & Terror In Firefighters

Fear Is	Terror Is
1. Crawling along in a dark, smoky hallway.	While looking for a pet python.
2. Being in a Haz Mat fire and not being able to read the labels.	Knowing what the labels mean.
3. Riding to a fire at breakneck speeds on a dark, icy night.	Being the driver.

Fear & Terror In Firefighters

Fear Is	Terror Is
4. Worrying about a floor collapsing.	On a suspension bridge.
5. Charging into a building to attack a fire.	Being left behind at the firehouse.
6. Venting a roof and not knowing what's going to pour out.	Working near a rookie whose using the new chain saw.

Handling The Media

Fires make great stories and the press and TV cameras are always going to be on hand. It's worthwhile using this exposure to further firefighting purposes (i.e., get more funding).

Help the media people share the thrill of firefighting with comments like...

1. "Yeah, this is a hot one, could blow at any time."
2. "If the gasses from those drums drift this way, you'll have only six seconds to live."
3. "If this wall collapses, of course, your tapes won't be worth a damn."

Handling The Media

The firefighters doing the really dangerous stuff won't get a bit of media attention. No one in their right mind would go in a burning building, and if they did, the smoke would prevent them from seeing anything. All the supporting guys, however, will have a great time waving to the family and kids. Every firefighter fantasizes about camera men catching his most dramatic rescue, but the guys with the cameras are never there when you really want them.

Fighting Fires & Burning Calories

The real hard part about fighting fires is that you have to do it while wearing this incredibly uncomfortable full turnout gear and carrying 60 pounds of SCBA and other equipment. It's learning to maneuver and climb, much less walk with all this stuff, that separates the rookies from the veterans. Firefighters burn-up so many calories lugging their loads up and down ladders and stairs, that they've got to eat lots of munchies at the firehouse to make up for the loss.

The guys at the firehouse had to constantly carbo load so they'd be sure to have energy in an emergency.

Munchies At The Firehouse

Hey, while you're waiting around for the next call-out, you've got to nibble on something to keep up your energy: chocolate covered and salty, roasted and dipped, spicy and crunchy, low in cholesterol (don't look at the fat content) and maybe even organically grown. No problem if there's no alarm for a while. There's that set of weights in the exercise room that you can work it off with as soon as the game on TV is over.

The Chief planted a trail of potato chips right to the exercise room.

Firefighters' Health Foods

There is no better way to maintain your energy and keep those hunger pangs away at the big emergency than by insuring you have at least one serving a day from the Firefighters' 5 Basic Food Groups.

5 Basic Firefighter Food Groups

1. Pizza
2. Donuts
3. Chocolate
4. Hamburgers
5. Diet Soda

Medical Emergencies

Administering first aid consists of the following:

1. Getting the victim to sign the release form.
2. Stopping the victim from screaming. Screaming upsets all the gapers as well as rookie EMT personnel.
3. Trying to help the victim so they will be able to fill out the rest of your forms and questionnaires, and make a contribution or buy a raffle ticket when there's a new piece of equipment you need.

"WAIT TILL HE'S BREATHING BEFORE YOU GET HIM TO SIGN THE RELEASES."

Blood

Firefighters and victims bleed a lot. What can I say? Firefighting is a dangerous job and victims just love to bleed which is why they are called victims. Other people's blood splashing about is bad enough, but there is nothing worse than when it's your own.

Blood can be in one of three places:

1. Inside of people where it belongs.
2. Splashing around on you and your clothing where it doesn't belong.
3. Being put in little glass vials at your annual physical.

If it is anywhere else, call a doctor.

"IF HE GIVES YOU ANY TROUBLE, SHOVE THIS GARLIC UP HIS NOSE."

Officers

Officers are important to firefighters. They take care of your morale (whatever that is) like when they explain why you have to drive the yellow tanker rather than a traditional red one, or why the department can't afford face masks you can actually see through. And they take care of your morals (whatever they are) like when they throw out everyone's good magazines and girlfriends from the back of the firehouse. Officers are best when there is a fire, lots of chaos and everyone running around, and then they assume command (whatever that is) and you put out the fire.

Cartoon Idea by Collie Grey

Firefighting Is 99% Hard Work

The other part is luck. Luck, like when it's not your shift when the scary cellar fire comes up or being sick the day little Johnny Livingston's pet alligator gets caught in the septic system.

"IT'S NOT EVEN MY REGULAR SHIFT."

Firefighters As Psychologists

Psychology is every bit as important as water in doing a firefighter's job. You have to pretend you know what you're doing when getting kids unstuck from someplace scary, and make believe you're medically competent so victims won't panic. When a TV camera shows up, you've got to act fearless and calm and competent or the citizens will cut you out of the budget next year. Firefighters are on the cutting edge of applied psychology every day.

"CENTERVILLE IS THE ONLY TOWN WHOSE FIREHOUSE HAS IT'S OWN PSYCHIATRIC WING."

Firefighters Know All Your Secrets

Visiting homes unexpectedly gives firefighters an incredibly intimate look at people's secret lives. Not only do they know about your burning the chuck roast in the oven you tried to wire yourself, or that you still keep your wallet under your mattress (how did the mattress catch on fire from one little cigarette?), but the bedroom collection of special interest magazines (if you get my drift) and X-rated videos are still keeping the guys glued to their VCR on rainy afternoons.

Firefighters Know All Your Secrets

The occasional fire at a gay club or after hours bar will give an entire department gag material for a decade. Professional firefighters try never to gossip about the leather, pantyhose, handcuffs and crotchless stuff they find in bedrooms. Likewise, they keep quiet about the identity of guests in hotel fires as well as who got locked in whose bathroom.

"RUN FOR YOUR LIFE! IT'S GONNA BLOW!"

Cartoon Idea By Collie Grey

Haz Mat

How are you supposed to know that cellulose decomposes to silicon dioxide when burned, which changes to sulfuric acid when it contacts water? Give me a break, you're firefighters, not chemical engineers. Or that freon decomposes to phosgene when it contacts a flame, which changes to hydrochloric acid when it contacts moisture. I mean there are some really scary substances in fires. Running the pumps may not have all the glory, but it doesn't get your kidneys and lungs dissolved in acid. For heavens sake, don't just sneak around the corner of a burning building for a little pee. That's liable to deteriorate into flavored vodka and the fumes will make all your buddies tipsy.

"GET READY BOYS! YOU'RE GOIN' IN!"

Cartoon Idea By Collie Grey

Firefighters' Cars

Firefighters' cars are not like most people's cars. They start on cold mornings and in wet weather. They usually have gas in the tanks and most parts work even if the finish is occasionally scorched. The back seats are good collecting places for junk. Most of the stuff there tends towards gloves, straps that you can't quite figure out what they go to and large wadded piles of training and instruction manuals. Under the seats are the gum wrappers, coins, maps and empty coffee cups, like in real people's cars.

"I THINK THE WIND'S SHIFTED."

Handling Hypochondriacs

A large part of firefighters' rescues involve hypochondriacs. These people have developed the game of worrying about themselves to a very high degree. A little indigestion: rush them to a cardiac unit. Bang a limb: an ambulance ride for an X-ray. Strain a back muscle moving lawn furniture: four firefighters have to carry them to a hospital. Hell, firefighters hurt their backs on a regular basis and they usually keep on working.

How To Grow Rich As A Firefighter

Anyone can write a book. Look at me. I can hardly spell and my wife still has to correct my grammar, and I write lots of books. If you want to grow rich effortlessly, this is the way. "What should I write about?" you ask. Write about firefighting. Write about what you find in people's bedrooms in that bottom night table drawer. Write about the videos they keep by the upstairs VCR. Write about the dumb places they lock themselves in and the ridiculous excuses they have for starting fires.

Brush Fires

Brush fires are nature's way of mowing the lawn. They are also nature's way of getting firefighters out into the country on sunny afternoons. When the wind changes, they can also be nature's way of sterilizing your car finish and replacing aging fire trucks with a fund raising campaign. Brush fires that grow in size can become full-fledged forest fires or even conflagrations. Brush fires give you the opportunity to run lines over highways and railroads and completely stop traffic, which can greatly expand the number of spectators cheering you on.

"WE'LL HAVE IT KNOCKED DOWN IN AN HOUR. WHY DO YOU ASK?"

Computers

Computer games and simulations are a great way to learn about fighting fires, and the Nintendo and Sega games are even better. They really are a lot more fun than checkers, even with its hundred year old tradition. Every department has a computer expert or two, usually younger firefighters, who can make the thing work. Older firefighters, and this includes most officers, have pretty much accepted that computers work mostly by magic.

Giving Directions

Giving directions is a bigger job at most firehouses than responding to emergencies. First, the lost motorist parks right in front of your doors and blocks all the equipment. Then the wife of the lost clown asks for directions since she's sure firefighters know where everything is in town. Men never ask. Men still have this macho thing about using maps and figuring it out for themselves which is a very important primeval instinct left over from caveman hunting days. Firefighters are actually the worst in this regard and may prefer to put out a fire in the wrong town rather than ask for directions.

Good And Bad Things To Bring To Fires

Good Things To Bring:	Bad Things To Bring:
1. Unfinished chicken leg or pizza from dinner.	1. The truck's air and hose lines.
2. Empty bladder.	2. Dalmatians.
3. Stout heart and clean soul.	3. Business cards from your painting & car repair business.
4. Saint Florin medallions and his bucket—if you can find it.	4. Walkman with tapes of sirens screaming & air horns blasting.

Mattress Fires

There was one emergency where a smoker fell asleep in a water bed, burned a hole in the mattress and drowned. In rural areas where people still save their money under their mattresses, more firefighters have been injured poking around trying to save the cash than you ever imagined. Fire resistant gloves were originally developed to handle this problem.

Firefighters' Vacations

Firefighters' vacations are the same as real people's vacations except that when they hear the sound of a siren or see a fire truck roar by, they drop their marshmallows or guidebook or crossword puzzles and rush to the scene. Come to think of it, firefighters rush to fires at home even when they are off-duty. There is something that pulls firefighters to fires that even "All-You-Can-Eat Buffets" can't keep away.

"OH, NO. WE HAD TO SEE HOW THEY HANDLE FIRES IN MEXICO."

Firefighters & Regularity

Being a firefighter is much harder on the gastrointestinal system than people realize. This is especially true with older firefighters whose bodies rebel at the constant interruptions to their toilet schedules. Just when Jake sits down in a contemplative mood for some serious business, an alarm is struck and he's off and running again. As likely as not, when the emergency is over, the intestines' interest in returning to business is also over, making the delay more or less permanent. Most laxatives were originally developed for firefighters.

"BRAN'S KICKING IN A LITTLE EARLY TODAY, HUH, BRAD?"

Traffic Preemption

I'm almost afraid to write that there are actually systems, using UHF radios, to turn lights green. Once the public learns how to break the digital codes (using home computers no doubt), these things are going to be more popular than radar detectors. Sure they'll outlaw them, but guys will hide them in the glove compartments and figure they'll have green lights for the rest of their lives. What'll happen when two cars with their gadgets approach the same intersection?

"WE JUST MADE IT."

Cartoon Idea by Collie Grey

Every Officer Has Their "Thing"

With some it's punctuality, with some it's neatness, some stress training, some cleanliness, some want initiative and some want you to only follow orders. Whatever their "thing," once you figure it out (and that won't take long), your life is going to be much easier and you'll have much more energy left for actually fighting fires.

"THE CHIEF HAS A THING ABOUT GERMS."

Haz Mat

The real problem with this "bad stuff" is the unpredictability of the goo. You never know if it's going to eat you, choke you, explode or just lie there and sag. Experienced firefighters know that in Haz Mat situations, it's best to let the officer go first, preferably the Chief, since he's the smartest, right?

"WAIT HERE WHILE I TIPPIE-TOE AROUND AND CHECK OUT THIS PLACARD NUMBER."

Cartoon Idea By Collie Grey

Stress Management

Stress for a firefighter may be defined as the confusion created by their mind when it overrides the body's basic desire to choke the living shit out of some civilian or politician who desperately deserves it. The other kind of stress is crawling along a hot, totally smoke-filled hallway when your SCBA starts sounding and you're not sure about the way out. Most firefighters can handle the latter.

Vultures

The vultures usually get to the fire a little before the firefighters and pretty much clog up the street, making it hard to get your equipment in place. The fire clean-up and restoration services are brawling on the street for first crack at the job; the lawyers are trying to look dignified as they pass out their cards in hope of 1/3 of any settlement; and the public insurance adjusters are trying to get the victim to sign contracts.

Vultures

The poor victim is pressed to make a hasty decision by the public adjuster and a typical conversation goes like this:

Victim at burning window: —Help, help!

Public Adjuster:—Good evening, Sir. I'm with the Adams Insurance Adjusting Company and we represent YOU in settlement negotiations with your insurance company.

Victim:—Help, help–I'm on fire!

Public Adjuster:—Yes, yes of course, and I'll bet you're suffering lots of mental anguish too–I might also suggest a lawyer friend.

Victim:—Help, help–Call the Fire Department!

Public Adjuster:—Certainly, Sir, if you'll just sign here and here, and initial pages one and two.

Firefighters' Farts

It's a fact. Firefighters can fart with the best of them. Burly construction workers, beer guzzling truck drivers, skinny little broccoli eaters, all don't stand a chance against a trained firefighter. For years, this propensity to gassiness was the reason most positions on fire trucks were open. It was found, however, that taxpayers were complaining about the odor as the fire equipment was passing through their neighborhoods, so nowadays, most trucks have enclosed places for the firefighters to ride. The required seat belts are more to keep the team members from bolting for the door because of the farts, than for any safety reasons.

"BILL THINKS HIS FARTING IS PERFECTLY ACCEPTABLE 'CAUSE HIS DOCTOR RECOMMENDED THE BROCCOLI."

Training

EMT
EMS
CPR
EAS
HAZ MAT
EME FIRST RESPONSE
ADVANCE EMT
EMERGENCY SERVICE
PARAMEDICS
FIRST AID SQUAD
EMT PARAMEDICS
AMBULANCE SQUAD

Actually, the life saving training is not so bad, it's getting all the titles, acronyms and abbreviations right that's the hard part. Who can figure out what all these people know and whose advice you should follow if they have a disagreement?

"I'M THE CHIEF... AND I GIVE MOUTH TO MOUTH."

Forcible Entry

You never have to worry about being locked out if there's a firefighter in your family or around as a neighbor. These guys can get into any place. They may scratch up the woodwork and crack the frame a bit, but you'll be in or out faster than with a key. It's a wonder the state penitentiary isn't filled up with firefighters who were trying to make a few bucks on the side.

"CHIEF SAYS WAIT FOR THE KEY."

Cartoon Idea by Collie Grey

Challenging Rescues

People get themselves into the weirdest situations. A guy in Boston fell out of a tree he was trimming. Fortunately, a safety belt saved him but it flipped him upside down and pulled his pants off leaving him hanging naked from the tree in freezing temperatures. I'm not even making this up, and most firefighters probably have better stories. There's material here for a great TV series though, of course, you'd want to use beautiful, big-breasted women to hang naked upside down.

The Firefighters' Commandments

I. Thou shalt remember to remove the air and power lines from thy truck before leaving thy firehouse.

II. Thou shalt not forget thy polypropylene underwear on cold nights.

III. Thou shalt heed thy SCBA warning bell or thou shalt become crispy.

IV. Thou shalt not smash into taxpayers' cars on the way to fires. It makes them reluctant to vote money for new equipment.

V. Thou shalt insure that all victims sign a release form.

The Firefighters' Commandments

VI. Thou shalt not gross out guests at parties with disgusting horror stories.

VII. Thou shalt let an officer go first when the HAZ MAT risk is real scary.

VIII. Thou shalt never accuse thy Chief of being a cheapskate if he is within hearing distance.

IX. Thou shalt be extra wary of electrical things 'cause no one understands this stuff.

X. Thou shalt not discuss thy intestinal problems while fellow firefighters are eating lunch.

These other books are available at many fine stores.

#2350 Sailing. Using the head at night • Sex & Sailing • Monsters in the Ice Chest • How to look nautical in bars and much more nautical nonsense.

#2351 Computers. Where computers really are made • How to understand computer manuals without reading them • Sell your old $2,000,000 computer for $60 • Why computers are always lonely and much more solid state computer humor.

#2352 Cats. Living with cat hair • The advantages of kitty litter • Cats that fart • How to tell if you've got a fat cat.

#2353 Tennis. Where do lost balls go • Winning the psychological game • Catching your breath • Perfecting wood shots.

#2354 Bowling. A book of bowling cartoons that covers: Score sheet cheaters • Boozers • Women who show off • Facing your team after a bad box and much more.

#2355 Parenting. Understanding the Tooth Fairy • 1000 ways to toilet train • Informers and tattle tales • Differences between little girls and little boys • And enough other information and laughs to make every parent wet their beds.

#2356 Fitness. T-shirts that will stop them from laughing at you • Earn big money with muscles • Sex and Fitness • Lose weight with laughter from this book.

#2357 Golf. Playing the psychological game • Going to the toilet in the rough • How to tell a real golfer • Some of the best golf cartoons ever printed.

#2358 Fishing. Handling 9" mosquitoes • Raising worms in your microwave oven • Neighborhood targets for fly casting practice • How to get on a first name basis with the Coast Guard plus even more.

#2359 Bathrooms. Why people love their bathroom • Great games to help pass the time on toilets • A frank discussion of bathroom odors • Plus lots of other stuff everyone out of diapers should know.

#2360 Biking. Why the wind is always against you • Why bike clothes are so tight • And lots of other stuff about what goes thunk, thunk, thunk when you pedal.

#2361 Running. How to "go" in the woods • Why running shoes cost more than sneakers • Keeping your lungs from bursting by letting the other guy talk.

#2362 Skiing. Understanding ski reports • Chair lift etiquette • Why trail maps don't show trees • Where moguls really come from • Rules for hot tubs and saunas.

#2363 Doctors. Handling lawyer and insurance problems with a rusty scalpel • Offshore medical schools and conferences • Why surgeons always get to carve the turkey on Thanksgiving • And a lot more humor that can be easily digested between patients.

#2364 Lawyers. Making faces at the judge • Why lawyers make better lovers • Quit law and make more money as a plumber • The first lawyer book ever written with more jokes for lawyers than about them.

#2365 Teachers. How teachers develop the Bladder of a Camel • Handling Back-to-School Nights and April Fools' Day • Recruiting an informer • Out-smarting kids who are smarter than you.

#2366 Nurses. Making the doctor look good • Ways to eject obnoxious visitors • Why nurses never get sick • Making big money in nursing • Great presents patients give you.

#2367 Firefighters. Why firefighters are irresistible to women • What firefighters do at parties • Rescuing gorillas out of trees • Saturday mornings at the firehouse.

#2368 Marines. How Marines handle stuffed up toilets • How Marines discipline their pets • Why the back seats of Marines' cars are so neat • Why Marines make better lovers and much more.

Ivory Tower Publishing Co., Inc. 125 Walnut St., PO Box 9132, Watertown, MA 02272-9132
Telephone #: (617) 923-1111 Fax #: (617) 923-8839